Bibliographic information published by the German National Library:

The German National Library lists this publication in the National Bibliography; detailed bibliographic data are available on the Internet at http://dnb.dnb.de .

Imprint:

Print and binding: Books on Demand GmbH, Norderstedt Germany
ISBN: 9783668947801

This book at GRIN:

https://www.grin.com/document/468970

Disha Kumari

General overview of medicinal plants

GRIN Verlag

Title

"General overview of medicinal plants"

CONTENTS

INTRODUCTION

Medicinal plants, also called medicinal herbs. Countries wealth resides in its plant inheritance. Researchers suggested that approx two out of three of world's plant species is medicinal. Most of the plants are found having high value of antioxidant potential. Antioxidants helps to reduce oxidative stress in the cell. This property is also helpful in treating many diseases like cancer, inflammatory diseases and cardiovascular diseases. The antioxidant potential are extracted from stems, roots, bark, fruits, seeds and leaves of several important medicinal plant. Synthetic antioxidants such as butylated hydroxytoluene (BHT) and butylated hydroxylanisole (BHA) are currently used as food additives, and many plant species have similar antioxidant potentials as these synthetics. The literature reveals that these natural antioxidants represent a potentially side effect-free alternative to synthetic antioxidants in the food processing industry. They are also used in preventive medicine. Medicinal plants have been discovered and used in traditional medicine practices since prehistoric times also hundreds of chemical compounds for functions including defence against insects, fungi, diseases, and herbivorous mammals were known since then. Numerous phytochemicals with potential or established biological activity have been identified and were researched with interest. A single plant contains widely diverse phytochemicals, the effects of using a whole plant as medicine are uncertain and needs a wide knowledge and understanding. Drug research discovered hundreds of useful compounds by the use of ethnobotany to search for pharmacologically active substances in nature. These contain the common drugs aspirin, digoxin, quinine, and opium. The compounds that are found in plants are of many kinds, but most are in four major biochemical classes as alkaloids, glycosides polyphenols, and terpenes

LITERATURE REVIEW

Chattopadhyay and Bhattacharrya, 2007 studied the *Terminalia chebula* (*Combretaceae*) is being used for the treatment of various types of diseases, because of its strong power of healing with a wide spectrum of biological activity. *Terminalia chebula* chemical constituents, has been achieved regarding the biological activity and medicinal applications of the herbal plant. It is a development of medicines and industrial products against diseases. It possess anti-oxidant activity, anti-diabetic activity, anti-inflammatory activity, anti-microbial, anti-pyretic, anti-tumor and anti-mutagenic activities etc. The bioactive components gallic acid, ellagic acid, tannins, chebulinic acid, chebulagic acid, bellericanin, rutin, quercetin etc are presence in the herbal plant.

Gowda *et al.*, 2012 reviewed on Triphala is one of the most powerful herbal drug used in medicine. Triphala is a combination of *Terminalia chebula*, *Phyllanthus emblica*, and *Terminalia belerica* species. It is also in the treatment of Acquired immune deficiency syndrome and various diseases. It contains many polyphenolic components, gallic acid, ascorbic acid (Vitamin C), ellagic acid, chebulinic acid, chebulagic acid, and flavonoids (quercetin and rutin). It shows the different pharmacological activities like anti-inflammatory activity, anti-cancer activity, anti-oxidant activity, anti-ulcer activity, anti-microbial activities etc.

Amit Kumar, 2014 states that Ayurveda is the oldest medical system. In the last some years the field of ayurveda herbal medicine and drugs are gaining positions in various countries. The medicinal herbs are shows natural origin and without side effects. The WHO has selected up to twenty one thousand herbal plants, which are used for medicinal purposes around the world. Mostly number of professional scientists discovered that one of the most value of all Ayurvedic combination - Triphala (*Terminalia chebula*, *Phyllanthus embilica* and *Terminalia bellirica*) in different ratios exhibits a number of health benefits, including various pharmacological activites like antioxidant, anticancer, antidiabetic etc.

Padmaa Paarakh, 2010 says that *Terminalia arjuna* (Roxb.) Wt. and Arn. (Arjuna; Combretaceae) is a widely used medicinal plant throughout India and popular in various Indigenous System of Medicine like Ayurveda, Siddha and Unani. In the Indian System of Medicine, the bark are used as astringent, cooling, tonic, aphrodisiac, cardiotonic, in fractures, ulcers, spermatorrhoea, leucorrhoea, diabetes, cough, tumour, excessive perspiration, asthma, inflammation and skin disorders etc.

Bariya Aditi Hemrajbhai et al., 2016 Medicinal plants have been a major source of therapeutic agents to cure human diseases, since ancient time. *Terminalia arjuna* is one kind of widely used medicinal plant used in various indigenous system of medicine like Ayurveda, Sidda and Unani. This review has been conducted to pile up phytochemical as well as pharmacological information that is available in different scientific literatures. This plant has been reported to contain phytochemical constituents like Triterpenoids (arjunolic acid, arjunic acid, arjunin, terminic acid, *etc.),* Glycosides (arjunetin, arjunosides, arjunglycosides etc.), Flavonoids (Arjunone, arjunolone etc.), Tannins (casuarinin, gallic acid, pyrocatechols etc.), β-Sitosterol, Minerals (calcium, magnesium, zinc, copper etc.). which exhibit various pharmacological activities like Antimicrobial, Anti cancer, Cardioprotective, Anti fungal, Anti diabetic, Antioxidant, Anti inflammatory, Hypolipidemic, Anthelmintic, Insecticidal, Wound healing, Anti acne, Gastroprotective etc. The present comprehensive review is therefore an effort to give detailed information on phytochemical and pharmacological studies of *T. arjuna*.

Anindita Deb et al., 2016 *Terminalia bellerica* Roxb (combretaceae) is found widely throughout the Indian subcontinent, Sri Lanka, South- East Asia, Bangladesh as a medicinal plant. Plant and plant parts are used in the traditional system of medicines like Ayurveda, Siddha, Unani & Chinese medicine. The plant is constituted of Glucoside, Tannins, Gallic acid, Ethyl Gallate, Chebulinic acid which serves as an antioxidant, antimicrobial, antidiarrheal, anticancer, antihypertensive, hepatoprotective& antipyretic agent. This review sites the information on pharmacological activities of *Terminalia bellerica* which may serve as a source for further research studies.

Ravi Kant Upadhyay (2016) Onion is a traditional nutraceutical and medicinal herbal plant.That is cultivated around the world. It contain phenolic compounds and flavonoids that havepotential anti-anaphylaxis, antioxidant, antiinflammatory, anti-cancer properties. Onions contain water (89%), protein (2%), and B1, B2, and C vitamins. It also contains polysaccharides such as saccharose, fructosans, quercetin essential oil and peptides. Onion contains numerous sulfur compounds (thiosulfinates and thiosulfonates) and sulfoxides. Onion is highly nutritional and its dietary use improves digestion and mental health and lower down toxigenicity of oils. Onion has potential in treating cardiovascular disease, hyperglycemia, and stomach cancer. Onion contains an important antioxidative, quercetin that is derived from *Allium cepa* on aldehyde oxidase lowdensity lipoprotein which reduces hepatocytes apoptosis in streptozotocin-induced diabetic rat. Onion has great ethno medicinal importance as native remedies used against diabetes, and related complications are from onion.

Guo-Qing Shi et al (2016) the anti-proliferative activity of flavonoids from onion skins, extraction by ethanol (50% v/v), soxhlet polar fractionation, thin-layer chromatography, pH gradient separation and recrystallization methods were used to isolate and purify flavonoids from dry onion skins. Anti-proliferative activities of some flavonoids obtained on leukemia K562 cell line were deter-mined by MTT assay method. Results showed that flavonoids of onion skins were mainly in form of quercetin, isorhamnetin, rutin, kaempferol, apigenin-7-O-Dglucopyranoside, quercetin-3-O-D-glucopyranoside, kaempferol-7-O-D-glucopyranoside. Kaempferol and Quercetin decreased K562 cell viability, and quercetin had stronger effect. However, rutin and isorhamnetin exhibited certain proliferation-promoting effects.

Olanrewaju Sam Olayeriju et al. (2015) The effects of ethylacetate extract of red onion (*Allium cepa*) tunic on some hemodynamic and biochemical parameters were evaluated in normotensive albino rats. Blood pressure parameters were determined in anaesthetized rats orally administered ACTE (10-, 20-, or 40 mg/kg) for 14 days. Respectively, 10-, 20-, or 40 mg/kg ACTE produced significant ($P < 0.01$), dose-dependent fall in systolic blood pressure, diastolic blood pressure, DBP (6%, 10%, 16%), pulse pressure, mean arterial blood pressure, HR (4%, 5%, 7%), MAPB (13%, 18%, 23%), SBP (21%, 27%, 33%) and PP (42%, 49%, 56%), and heart rate. The highest effective dose (40 mg/kg) compared well with ramipril (1 mg/kg) with regards to DBP (19%), SBP (41%), HR (10%), MABP (29%) and PP (70%). Similar trends (decreases) were recorded for 40 mg/kg ACTE and ramipril, respectively, as regards the activities of serum enzymes: ALT (18% and 14%), creatine kinase (60% and 65%) and ALP (28% and 16%). High Pressure Liquid Chromatography fingerprints of the flavonoid-rich ACTE revealed that rutin, quercitrin, kaempferol, and isoquercitrin are the active flavonoids.

Lee J and Mitchell AE (2011). In the following study, the flavonoid composition was determined in coproduct materials and the press cake (material generated after juice extraction) in several commercially important onion varieties. Bio flavonoids were characterized and quantified using LC-(ESI) MS/MS and HPLC analysis. The long-term stability of quercetin glycosides was assessed in dried coproduct materials stored at 5 and 21 °C over one year. In all varieties, the predominant forms of quercetin were the quercetin and 4'-O-glucoside.quercetin 3,4'-O-glucoside. Onoin contained the highest levels of flavonoids (1613 mg/100 g on a dry weight basis (DW).

Praveen Kumar A and Bhawana S. (2013) Ginkgo biloba L., a plant belongs to the Ginkgoaceae family. Rutin is that the major flavonoid glycoside found in gingko that has been referred to as vitamin P. The isolation of rutin by completely done with chromatographically methodology. The isolated rutin was known by measure its FTIR spectra, HPLC, TLC, melting point and ultraviolet absorption. It shows various pharmacological effects and controls the various diseases.

Maria Isabel Cardona et al (2017) *Physalis peruviana* is an exotic fruit highly valued for its bioactive compounds. This work evaluates the effect of some variables of the extraction process of *P. peruviana* calyces extract on total phenolic content (tpc), total flavonoid content, and in vitro antioxidant activity of the extracts obtained. The response variables were then evaluated using in vitro assays such as NO and DPPH uptake for antioxidant activity; the Folin-Ciocalteu method for total phenolic content; and the external standard method by HPLC-DAD for quantification of flavonoids.

Fatemeh Fathiazada et al (2006) the potential of tobacco waste as a viable source of flavonoids like quercetin and rutin has been investigated. These flavonoids have been isolated from waste tobacco leaves, and their identities have been confirmed by 1H-NMR, 13C-NMR spectroscopy and UV-visible. The amount of rutin present in the tobacco leaves by using HPLC analysis.

Stevenson DG et al (2017) Twelve pumpkin cultivars (Cucurbita maxima D.), cultivated in Iowa, were studied for their seed oil content, fatty acid composition, and tocopherol content. Oil content ranged from 10.9 to 30.9%. Total unsaturated fatty acid content ranged from 73.1 to 80.5%. The predominant fatty acids present were linoleic, oleic, palmitic, and stearic. Significant differences were observed among the cultivars for stearic, oleic, linoleic, and gadoleic acid content of oil. Low linolenic acid levels were observed (<1%). The tocopherol content of the oils ranged from 27.1 to 75.1 microg/g of oil for alpha-tocopherol, from 74.9 to 492.8 microg/g for gamma-tocopherol, and from 35.3 to 1109.7 microg/g for delta-tocopherol. The study showed potential for pumpkin seed oil from all 12 cultivars to have high oxidative stability that would be suitable for food and industrial applications, as well as high unsaturation and tocopherol content that could potentially improve the nutrition of human diets.

Li T et al (2012) Taeniasis refers to the infection with adult tapeworms of Taenia spp. In the upper small intestine of humans, which is also a cause of cysticercosis infection in either both humans and/or animals. Currently the most commonly applied anthelminthics for treatment of taeniasis are praziquantel and niclosamide. Praziquantel is very effective, but has the risk of induction of epileptic seizures or convulsions in carriers with asymptomatic concurrent neurocysticercosis. In contrast, niclosamide is safe and effective, but is not readily available in many endemic countries including China. In the current community-based study, we assessed the curative effect of either pumpkin seeds or areca nut extract alone in taeniasis, and also looked at synergistic effects of these two herb drugs on tapeworms. In the study group with the pumpkin seed/areca nut extract treatment, 91 (79.1%) of 115 suspected taeniasis cases (with a history of expulsion of proglottids within the previous one year) released whole tapeworms, four (3.5%) expelled incomplete strobila, and no tapeworms or proglottids were recovered in the remaining

20cases. In these 115 persons, 45 were confirmed as taeniasis before treatment by microscopy and/or coproPCR. Forty (88.9%) of 45 confirmed cases eliminated intact worms following treatment. The mean time period for complete elimination of tapeworms in 91 taeniasis cases was 2 h (range 20 min to 8 h 30 min), and 89.0% (81) of 91 patients discharged intact worms within 3h after drug administration. In Control Group A with treatment of pumpkin seeds alone, 75.0% (9/12) of confirmed taeniasis cases expelled whole tapeworms, but the mean time period for complete elimination was about 14 h 10 min (range 3 h 20 min to 21 h 20 min), which was much longer than that (2h) for the study group, whereas in Control Group B treated with areca nut extract alone, only 63.6% (7/11) of taeniasis cases discharged whole tapeworms, and the mean time period was 6 h 27 min (range 1-22 h). Mild side effects included nausea and dizziness in about 46.3% of patients with the pumpkin seeds/areca nut extract treatment, but all discomforts were transient and well tolerated. In conclusion, a synergistic effect of pumpkin seed and areca nut extract on Taenia spp. tapeworms was confirmed in the current study, primarily in producing an increased rate of effect on tapeworm expulsion (average time 2 h for combination vs 6-21 h for individual extracts). The pumpkin seed/areca combined treatment was indicated to be safe and highly effective (89%) for human taeniasis.

Caili F et al (2006) Dietary plants and herbal preparations have been traditionally used as medicine in developing countries and obtained a resurgence of use in the United States and Europe. Research carried out in last few decades has validated several such claims of use of traditional medicine plants. Popularity of pumpkin in various systems of traditional medicine for several ailments (antidiabetic, antihypertensive, antitumor, immunomodulation, antibacterial, antihypercholesterolemia, intestinal antiparasitia, antiinflammation, antalgic) focused the investigators' attention on this plant. Considerable evidence from several epidemiological studies concerning bioactivities leads have stimulated a number of animal model, cell culture studies and clinical trials designed to test this pharmacological actions. In addition, it was found that technologies such as germination and fermentation could reduce antinutritional materials and affect the pharmacological activities of pumpkin. This review will focus on the main medicinal properties and technologies of pumpkin, and point out areas for future research to further elucidate mechanisms whereby this compound may reduce disease risk.

FU S et al (2006) Dietary plants and herbal preparations have been traditionally used as medicine in developing countries and obtained a resurgence of use in the United States and Europe. Research carried out in last few decades has validated several such claims of use of traditional medicine plants. Popularity of pumpkin in various systems of traditional medicine for several ailments (antidiabetic, antihypertensive, antitumor, immunomodulation, antibacterial, antihypercholesterolemia, intestinal antiparasitia, antiinflammation, antalgic) focused the investigators' attention on this plant. Considerable evidence from several epidemiological studies concerning bioactivities leads have stimulated a number of animal model, cell culture studies and clinical trials designed to test this pharmacological actions. In addition, it was found that technologies such as germination and fermentation could reduce antinutritional materials and affect the pharmacological activities of pumpkin. This review will focus on the main medicinal properties and technologies of pumpkin, and point out areas for future research to further elucidate mechanisms whereby this compound may reduce disease risk.

Anwesa Bag et al (2013) explained the medicinal plants are part and parcel of human society to combat diseases from the dawn of civilization. *Terminaliachebula* Retz. (Fam. *Combretaceae*), is called the 'King of Medicine' in Tibet and is always listed at the top of the list of 'Ayurvedic Materia Medica' because of its extraordinary power of healing. The whole plant possesses high medicinal value and traditionally used for the treatment of various ailments for human beings. Some of the folklore people used this plant in the treatment of asthma, sore throat, vomiting, hiccough, diarrhea, dysentery, bleeding piles, ulcers, gout, heart and bladder diseases. The plant has been demonstrated to possess multiple pharmacological and medicinal activities, such as antioxidant, antimicrobial, antidiabetic, hepatoprotective, anti-inflammatory, antimutagenic, antiproliferative, radio protective, cardio protective, antiarthritic, anticaries, gastrointestinal motility and wound healing activity. But no systematic updated information on the therapeutic effectiveness of *Terminalia chebula*, a popular herbal remedy in India and South-East Asia has so far been reported. This review highlights an updated information particularly on the photochemistry and various pharmacological and medicinal properties of *Terminalia chebula* Retz. and some of its isolated compounds, along with their safety evaluation. This may provide incentive for proper evaluation of the plant as medicinal agent against the human diseases and also to bridge the lacunae in the existing literature and future scope which may offer immense opportunity for researchers engaged in validation of the traditional claims and development of safe and effective botanical medicine.

Mishra (2011) explained that the Rasayan branch of Ayurveda deals specifically with and Rasayan formulations bestows upon the user, the longevity, with age stabilization, retaining youth for longer with maintaining strength of all organs optimally, enhanced intelligence, aphrodisiac prowess, improved complexion, voice and allied positive health attributes. Presently considerable research is being carried out on various Rasayan Products and the herbs to screen them for various therapeutic benefits. But the research is focusing only one or couple of aspects of Rasayan's therapeutic benefits. The exhaustive research on all therapeutic benefits of single Rasayan product will be more conclusive validation of claims of classical Ayurvedic texts. This perspective review also details main Rasayan formulations with emphasis on the herbs used in them, so that we could direct our focus of research on them and that too exhaustively.

Sharma et al (2013) observed that natural compounds from plants and other life forms (bacteria, fungi, marine organisms) represent a major source of molecules with medicinal properties. Among them, antioxidant substances are of particular interest. The understanding of the central role that oxidative stress holds in the progression of disorders as varied as: cardiovascular diseases, degenerative conditions, rheumatic disorders, metabolic syndrome, and in aging, makes antioxidant capacity to a key-feature of modern, multipotent remedies. A lot of medicinal plants, traditionally used for thousands of years, are present in a group of herbal preparations of the Indian traditional health care system (Ayurveda) named Rasayana proposed for their interestingantioxidant activities.

Prasan Bhandari and Mohammad Ameeruddin Kamdod (2012) studied the *Emblica officinalis Gaertn.* or *Phyllanthus emblica* Linn, commonly known as Indian gooseberry or Amla, is perhaps the most important medicinal plant in the Indian traditional system of medicine, the Ayurveda. Several parts of the plant are used to treat a variety of diseases, but the most important is the fruit. Many ailments are treated by the fruit which is used either alone or in combination with other plants. These include common cold and fever; as a diuretic, laxative, liver tonic, refrigerant, stomachic, restorative, alterative, antipyretic, anti-inflammatory, hair tonic; to prevent peptic ulcer and dyspepsia, and as a digestive. *E. officinalis* possesses antipyretic, analgesic, antitussive, antiatherogenic, adaptogenic, cardioprotective, gastroprotective, antianemic, antihypercholesterolemic, wound healing, antidiarrheal, antiatherosclerotic, hepatoprotective, nephroprotective, and neuroprotective properties as demonstrated in numerous preclinical studies. Furthermore, experimental studies have reported that *E. officinalis* and some of its phytochemicals also exhibit anticarcinogenic properties. E.officinalis is also reported to possess radiomodulatory, chemomodulatory, chemopreventive, free radical scavenging, antioxidant, anti-inflammatory, antimutagenic and immunomodulatory activities. These properties are efficacious in the treatment and prevention of cancer. This review summarizes the results related to these properties and also emphasizes the aspects that warrant future research establishing its activity and utility as a cancer preventive and therapeutic drug in humans.

Patel and Goyal (2012) informed that *Emblica officinalis* Geart. fruit is an important dietary source and its use as a medicinal plant has increased many fold over a period of time. A survey of the literature relating to the medicinal uses of *E. officinalis* fruit reveals it to be one of the widely used medicinal plants. It finds place in folklore medicine, Ayurveda, Unani and other indigenous systems of medicine. Following a large number of claims on the wide range of traditional medicinal properties of the plant, considerable efforts have been made to verify its efficacy as a curative agent through pharmacological investigations. In this study, a comprehensive account of the phytochemical investigation and therapeutic potential of *Embilica officinalis* fruit is presented.

Eugin Amala and Jeyaraj (2015) informed that the *Terminalia chebula, Terminalia bellirica, Phyllanthus emblica* and their formulations are used as a phytotherapeutic medicine for hepatic diseases. The effect of methanolic extract of herbal powder was administered against Imidacloprid in Albino Wistar Rats. Single dose of herbal (500 mg/kg/b.w) was given orally for 28 days to Imidacloprid toxicity exposed to rats. They were observed each hour and each day for 28 days for any changes in behavioural activity loss of ability to move, cramping, lethargy, muscle weakness, convulsion, irritation of the eye, excess salivation, change in gait. Clinical observations body weight, organ weight, biochemical analysis and histopathological examination were carried out. Administration of Imidacloprid, produced liver degeneration such as cytoplasmic vacuolation, mild vacuolar degeneration, mild hepatic damage in low dose (40 mg/kg b.w) and appearance of blood streaks, hepatic damage and severe vacuolar degenerations were observed in high dose of induction (80 mg/kg b.w). Animals were treated with the phytotherapeutic products of *Terminalia chebula, Terminalia bellirica, Phyllanthus emblica* and their formulations. Triphala extract was possessed more hepatoprotective activity, while comparing the four samples which was proved experimentally.

Bao Yang et al (2011) explained the mature longan (*Dimocarpus longan* Lour.) fruit has a succulent, edible and white aril, which has gained popularity as an exotic fruit in temperate regions. It is prized on the international market resulting in an increased production with significant contributions to local economic development. Longan fruit contains significant amounts of bioactive compounds such as corilagin, ellagic acid and its conjugates, 4-*O*methylgallic acid, flavone glycosides, glycosides of quercetin and kaempferol, ethyl gallate 1-*β*-*O*-galloyl-*d*-glucopyranose, grevifolin and 4-*O*-*α*-*l*-rhamnopyranosyl-ellagic acid. The fruit has been used in the traditional Chinese medicinal formulation, serving as an agent in relief of neural pain and swelling. The application of ultrasonic-assisted extraction or high pressure-assisted extraction greatly increases the yield from longan pericarp or seeds. In recent years, some pharmacological activities such as anti-tyrosinase, anti-glycated and anticancer activities, and memory-enhancing effects of longan aril, pericarp or seed extract have been found, implicating a significant contribution to human health. Regarding the increasing cultivation area and increasing quantity of longan fruit in the world, further utilization of this fruit is expected in an effort to use more efficiently the inherent bioactive compounds. The paper reviews the recent advances in the extraction and pharmacological activities of bioactive compounds from longan fruit. Some novel pharmacological potential of longan fruit is also discussed in this paper.

Naveen Srinivas et al (2016) says that Tulsi as a medicine," "Tulsi as a medicinal plant," "Medicinal properties of *Ocimum sanctum*," and "Tulsi in dentistry." The search was limited to articles and books in the English literature. To prepare a thorough review on the therapeutic aspects of Tulsi, the contents were screened between the year 1985 to 2015 by going through the title and abstracts, and further shortlisting articles for full text reading.

U Anyaehie (2009) The global scenario is now supporting the development of modern drugs from less toxic plant products with proven medicinal properties. Each part of neem plant (Azadirachta indica A. Juss) reportedly has various medicinal properties and has been in use in many continents for centuries. Recently, a fractionated neem-leaf extract known as IRAB with reported activities against Malaria, HIV/AIDS and cancer has been developed into a drug and currently marketed in Nigeria as IRACAP®. This paper reviews the medicinal properties, clinical studies and safety concerns of this fractionated acetone-water neem leaves extract as a footstep to further studies both on the extract and/or its chemical constituents.

Douglas Grindlay and T.Reynolds (2004) explained The mucilaginous gel from the parenchymatous cells in the leaf pulp of *Aloe vera* has been used since early times for a host of curative purposes. This gel should be distinguished clearly from the bitter yellow exudate originating from the bundle sheath cells, which is used for its purgative effects. *Aloe vera*gel has come to play a prominent role as a contemporary folk remedy, and numerous optimistic, and in some cases extravagant, claims have been made for its medicinal properties. Modern clinical use of the gel began in the 1930s, with reports of successful treatment of X-ray and radium burns, which led to further experimental studies using laboratory animals in the following decades. The reports of these experiments and the numerous favourable case histories did not give conclusive evidence, since although positive results were usually described, much of the work suffered from poor experimental design and insufficiently large test samples. In addition some conflicting or inconsistent results were obtained. With me recent resurgence of interest in *Aloe vera* gel, however, new experimental work has indicated the possibility of distinct physiological effects. Chemical analysis has shown the gel to contain various carbohydrate polymers, notably either glucomannans or pectic acid, along with a range of other organic and inorganic components. Although many physiological properties of the gel have been described, there is no certain correlation between these and the identified gel components.

Matan et al (2006) Mixtures of cinnamon and clove oils were tested for inhibitory activity against important spoilage microorganism of intermediate moisture foods. Four fungal species (*Aspergillus flavus*, *Penicillium roqueforti*, *Mucor plumbeus* and *Eurotium* sp.), four yeasts species (*Debaryomyces hansenii*, *Pichia membranaefaciens*, *Zygosaccharomyces rouxii* and *Candida lipolytica*), and two bacteria species (*Staphylococcus aureus* and *Pediococcus halophilus*) inoculated separately on agar plates were sealed in a barrier pouch and exposed to essential oil volatiles under a modified at mosphere of low O_2 (< 0.05–10%) and high CO_2 (20% or 40%), with the balance being N_2. *A. flavus* and *Eurotium*sp. proved to be the most resistant microorganisms. Cinnamon and clove oils added between 1000 and 4000 μL at a ratio of 1:1 were tested for minimum inhibitory volume (MIV) against molds and yeasts. The gas phase above 1000 μL of the oil mixture inhibited growth of *C. lipolytica* and *P. membranaefaciens*; 2000 μL inhibited growth of *A. flavus, P. roqueforti*, *M. plumbeus,*

Eurotium sp., *D. hansenii,* and *Z. rouxii*, while inhibition of *A. flavus* required the addition of 4000 μL. Higher ratios of cinnamon oil/clove oil were more effective for inhibiting the growth of *A. flavus*.

Shrikant Mishra and Kalpana Palanivelu (2008) says the effects of curcumin on patients with Alzheimer's disease (AD). Curcumin (Turmeric), an ancient Indian herb used in curry powder, has been extensively studied in modern medicine and Indian systems of medicine for the treatment of various medical conditions, including cystic fibrosis, haemorrhoids, gastric ulcer, colon cancer, breast cancer, atherosclerosis, liver diseases and arthritis. It has been used in various types of treatments for dementia and traumatic brain injury. Curcumin also has a potential role in the prevention and treatment of AD. Curcumin as an antioxidant, anti-inflammatory and lipophilic action improves the cognitive functions in patients with AD. A growing body of evidence indicates that oxidative stress, free radicals, beta amyloid, cerebral deregulation caused by bio-metal toxicity and abnormal inflammatory reactions contribute to the key event in Alzheimer's disease pathology. Due to various effects of curcumin, such as decreased Beta-amyloid plaques, delayed degradation of neurons, metal-chelation, anti-inflammatory, antioxidant and decreased microglia formation, the overall memory in patients with AD has improved. This paper reviews the various mechanisms of actions of curcumin in AD and pathology.

Benjamin H.S et al (1983) explain the effects of garlic on lipid metabolism and atherosclerosis. In both animal and human studies, components of garlic have been shown to lower cholesterol and triglyceride levels. Garlic has also been shown to change blood lipoproteins and to affect coagulation parameters. The available data suggest that garlic may be of value in either the prevention or treatment of atherosclerotic diseases. Further research in this area is warranted.

LIST OF PLANT SPECIES

S.No	***Botanical Name***	**Family**	**Habit**	**Medicinal Uses**
1	*Achyranthes aspera*	Amaranthaceae	Herb	Piles;Dieretic
2	*Actinopteris radiata*	Actinopteridaceae	Herb	Skin diseases
3	*Adiantum caudatum*L.	Adiantaceae	Herb	Diabetes
4	*Adiantum lunulatum*	Adiantaceae	Herb	Ulcers;Dysentery
5	*Aerva lantana*	Amaranthaceae	Herb	Urinary diseases
6	*Ageratum conyzoides*	Asteraceae	Herb	Nervine tonic
7	*Burm*	Acanthaceae	Herb	Fever
8	*Androghapisnallamlayana*	Acanthaceae	Herb	Leucorrhoea
9	*Argemeonemexicana*	Pepepaveraceae	Herb	Syphillis
10	*.Aristolochiabraceteolata*	Aristolochiaceae	Herb	Skin diseases; Diabetes
11	*Aristolochiaindica*	Aristolochiaceae	Herb	Snake bite
12	*Arisaematortuosum*	Araceae	Herb	Kill worms in Cattle
13	*Asperagusgonclados*	Liliaceae	Herb	Galactagogue;Diabetes
14	*Bacopa monnieri*	Scrophulriaceae	Herb	Memory booster
15	*Blumeamollis*	Asteraceae	Herb	Dropsy; Throat Infection
16	*Boerhaviadiffusa*	Nyctaginaceae	Herb	Urinary disorders
17	*Canscoradiffusa*	Gentianaceae	Herb	Piles; Skin diseases
18	*Cassia absus*	Caesalpinaceae	Herb	Constipation; Cough
19	*Cassia italica*	Caesalpinaceae	Herb	Bone fracture
20	*Centellaasiatica*	Apiaceae	Herb	Brain tonic

21	*Celosia argentea*	Amaranthaceae	Herb	Eczema;Ulcers
22	*Chlorophytumarundinaceum*	Liliaceae	Herb	Aphrodisiac;Hydrocele
23	*Cleome viscosa*	Cleomaceae	Herb	Wounds
24	*Cissusquadrangularis*	Vitaceae	Herb	Vitaceae
25	*Coldeniaprocumbens*	Boraginaceae	Herb	Rheumatism
26	*Costusspeciosus*	Costaceae	Herb	Snake bite
27	*Curculigoorchioides*	Hypoxidaceae	Herb	Asthma;Anaemia
28	*Cymopogonflexuosus*	Poaceae	Herb	Citral oil
29	*Datura metal*	*Datura metal*	Herb	Epilepsy
30	*Desmodiumtriflorum*	Fabaceae	Herb	Galactagogue;Dysentery
31	*Ecliptaprostrata*	Asteraceae	Herb	Asthma; Jaundice
32	*Enicostemaaxillare*	Gentianaceae	Herb	Scabies;Gout
33	*Euphorbia hirta*	Euphorbiaceae	Herb	Cough;Dysentery
34	*Evolvulusalsinoides*	Convolvulaceae	Herb	Asthma;Fevers
35	*Hedtituspeberula*	Rubiaceae	Herb	Asthma;bronchites
36	*Hedyotisaffinis*	Rubiaceae	Herb	Gonorrhoea
37	*Hedyotiscorymbosa*(L.)	Rubiaceae	Herb	Stomachic;diarrhoea
38	*Hedyotisherbacea*L.	Rubiaceae	Herb	Febrifuge,Rheumatism
39	*Hemidesmus indicus*	Periplocaceae	Herb	Cardio tonic
40	*Hemionitisarifolia*	Hemionitidaceae	Herb	Antidiabetes
41	*Hybanthusenneaspermus*	Violaceae	Herb	Impotency;Antidote
42	*Leonotisnepetiifolia*	Lamiaceae	Herb	Burns; Rheumatism

43	*Lepidogathiscristata*	Acanthaceae	Herb	Dandruf;Dental problems
44	*Limnophilaindica*	Scrophulriaceae	Herb	Dysentery;Carminative
45	*Pavoniaxylanica*	Malvaceae	Herb	Antihelmenthic;Febrifuge
46	*Peristrophepaniculata*	Acanthaceae	Herb	Wounds;Skin diseases
47	*Phyllanthus amarus*	Euphorbiaceae	Herb	Jaundice
48	*Abulitonindicum*	Malvaceae	Shrub	Haematuria
49	*Aegle marmelos*	Rutaceae	Shrub	Dysentery
50	*Anisomeleamalabarica*	Lamiaceae	Shrub	Diarrohoea;Eczema
51	*Atalantiamonophylla*	Rutaceae	Shrub	Fever; Antiseptic
52	*Azimatetracantha*Lam.	Salvadoraceae	Shrub	Leprosy; Eczema
53	*Baliospermummontanum*	Euphorbiaceae	Shrub	Stomach Pain;Laxative
54	*Capparis sepiaria*	Capparaceae	Shrub	Febrifuge
55	*.Carmonaretusa*	Boraginaceae	Shrub	Snake bite; Sterility
56	*Cassia auriculata*	Caesalpiniaceae	Shrub	Diabetes
57	*Cassia montana*	Caesalpinaceae	Shrub	Body Pains
58	*Cassia occidentalis*	Caesalpinaceae	Shrub	Laxative
59	*Catunaregam spinosa*	Rubiaceae	Shrub	Astringent;Dysentery
60	*Cipadessabaccefera*	Meliaceae	Shrub	Diabetes;Wounds
61	*Cissampelospareira*	Menispermaceae	Shrub	Dropsy;Diabetes
62	*Clerodendrumserratum*	Veranaceae	Shrub	Febrifuge;Diabetes
63	*Datura stromonium*	Solanaceae	Shrub	Asthma;Narcotic
64	*Decalepishamiltonii*	Asclepediaceae	Shrub	Haemorrhage;Appetizer
65	*Decaschistiacrotonifolia*	Malvaceae	Shrub	Hydrocele

66	*Desmodiumpulchellum*	Papilionaceae	Shrub	Diarrhoea;Aphrodisiac
67	*Dodoneaviscosa*	Sapindaceae	Shrub	Bone fractures; Antipyretic
68	*Euphorbia nivulia*	Euphorbiaceae	Shrub	Ruefacient;Diabetes
69	*Ficushispida*	Moraceae	Shrub	Leucoderma
70	*Flacouriaindica*	Flacourtiaceae	Shrub	Astringent;Appetiser
71	*Glycosmis pentaphylla*	Rutaceae	Shrub	Diabetes; Eczema
72	*Gymnemasylvestre*	Asclepediaceae	Shrub	Diabetes
73	*Helictersisora*	Sterculiaceae	Shrub	Desentery;Diabetes
74	*Indigoferacassioides*	Papilionaceae	shrub	Cough
75	*Justicia adhatoda*	Acanthaceae	Shrub	Asthma; Diabetes
76	*Lantana camara*	Verbanaceae	Shrub	Cuts; Wounds
77	*Maeruvaoblongifolia*	Capparidaceae	Shrub	Diabetes;Stimulant
78	*Ochnaobtusata*	Ochnaceae	Shrub	Asthma
79	*Paracalyxscariosa*	Papilionaceae	Shrub	Dysentery;Astringent
80	*Acacia leucophloea*	Mimosaceae	Tree	Diuretic
81	*Accacianilotica*	Mimosaceae	Tree	Cough;Diarrhoea
82	*Alangiumsalvifollium*	Alangiaceae	Tree	Dog bite
83	*Albiziaamara*	Mimosaceae	Tree	Inflamation
84	*Albiziaodoratissima*	Mimosaceae	Tree	Leprosy
85	*Annona reticulata*	Annonaceae	Tree	Astringent
86	*Anogeissuslatifolia*	Combretaceae	Tree	Bitter tonic;Astring
87	*Azadirachtaindica*	Meliaceae	Tree	Skin diseases
88	*Bauhinia racemosa*	Caesalpinaceae	Tree	Malaria fever

89	*Boswelia serrata*	Burseraceae	Tree	Arthritis
90	*Buchananialanzam*	Anacardiaceae	Tree	Leucorrhoea
91	*Buchananiaoxilaris*	Anacardiaceae	Tree	Cardio tonic; Wounds
92	*Caeseariaelliptica*	Flacourtiaceae	Tree	Diabetes;Diuretic
93	*Careyaarborea*	Barringtoniaceae	Tree	Eye diseases; Skin sores
94	*Canthiumdicocum*	Rubiaceae	Tree	Fever;Diarrhoea
95	*Cassia fistula*	Caesalpinaceae	Tree	Diabetes
96	*Cassine glauca*	Celastraceae	Tree	Eczema;Antidote
97	*Celastruspaniculatus*	Celastraceae	Tree	Brain tonic;Scabies
98	*Chloroxylonswieteniar*.	Meliaceae	Tree	Rheumatism; Astringent
99	*Cochlospermumreligiosum*	Cochlospermaceae	Tree	Dysentery;Gonorrhoea
100	*Commiferacaudata*	Burseraceae	Tree	Rheumatism
101	*Cordia dichotoma*	Boraginaceae	Tree	Bronchial disorders; Fever
102	*Dalbergialatifolia*	Fabaceae	Tree	Leprosy;Ulcers
103	*Dalbergiapaniculata*	Fabaceae	Tree	Filarial Swelling
104	*Deccanniapubscens*	Rubiaceae	Tree	Sores
105	*Dilleniapentagyna*	Dilleniaceae	Tree	Cooling agent for cattle
106	*Diospyros ebenum*	Ebenaceae	Tree	Astringent
107	*Diospyosmelanoxylon*	Ebenacaea	Tree	Astringent;diuretic
108	*Erithroxylonmonogynum*	Erythroxylaceae	Tree	Diuretic;Febrifuge
109	*Ficusbenghalensis*	Moraceae	Tree	Rheumatism
110	*Ficusmicrocarpa*	Moraceae	Tree	Wounds;Diabetes
111	*Ficustinctoria*	Moraceae	Tree	Fever;Stringent

112	*Gardenia resinifera*	Rubiaceae	Tree	Constipation; Bronchitis
113	*Givotiamoluccaina*	Euphorbiaceae	Tree	Drandruf;Psoriasis
114	*Gyrocarpusamericanus*	Hernandiaceae	Tree	Cancer
115	*Gyrocarpusasiaticus*	Hernandiaceae	Tree	Cancer
116	*Haldiniacordifolia*	Rubiaceae	Tree	Stomachic
117	*Hardwickiabinata*	Caesalpiniaceae	Tree	Rheumatism
118	*Holarrhenaantidysenterica*	Apocynaceae	Tree	Diabetes;Antidysentric
119	*Holarrhenapubescens*	Apocynaceae	Tree	Mouth Ulcers;Diabetes
120	*Holopteleaintegrifolia*	Ulmaceae	Tree	Piles;fistula;Scabies
121	*Ixora arborea*	Rubiaceae	Tree	Dysentery
122	*Ixora pavetta*	Rubiaceae	Tree	Urinary disorders
123	*Lanneacoromandalica*	Anacardiaceae	Tree	Nervous diseases; Diabetes
124	*Lepisanthestetraphylla*	Sapindaceae	Tree	Skin diseases;Dandruf
125	*Litseaglutinosa*	Lauracea	Tree	Bone fracture
126	*Madhucalongifolia*	Sapotaceae	Tree	Diabetes;Vigour
127	*Manikarahexandra*	Sapotaceae	Tree	Aphrodisiac; Ulcers
128	*Maytenusemarginata*	Caelastraceae	Tree	Gastrointestinal problems
129	*Oroxylumindicum*	Bignoniaceae	Tree	Rheumatism;Oedema
130	*Phyllanthus embllica*	Euphorbiaceae	Tree	Asthma
131	*Alstoniascholaris*	Apocynaceae	Climber	Abdominal pains
132	*Cardiospermumhalicacabum*	Sapindaceae	Climber	Rhematism
133	*Cassythafiliformis*	Lauraceae	Climber	Hair tonic; Dysentery

134	*Cissusvetigenia*	Vitaceae	Climber	Wounds
135	*Cocculushirsutus*	Menispermaceae	Climber	Rheumatism
136	*Ctenolepisgracini*	Cucurbitaceae	Climber	Boils;Wounds
137	*Derris scandens*	Fabaceae	Climber	Snake bite
138	*Gloriosa superba*	Liliaceae	Climber	Abortifacient;Leprosy
139	*Hiptagebenghalensis*	Malphigiaceae	Climber	Liver disorders;Dysentery
140	*Holostemmaada-kodein*	Asclepediaceae	Climber	Diabetes;Gonorrhoea
141	*Ichnocarpusfrutescens*	Apocynaceae	Climber	Blood Purifier
142	*Ipomeamauritiana*jacq	Convolvulaceae	Climber	Galactogogue
143	*Jasminum auriculatum*	Oleaceae	Climber	Dropsy;Expectorent
144	*Lygodiumflexuosum*	Schizaeaceae	Climber	Scabies;Eczema
145	*Momordica charantia*	Cucurbitaceae	Climber	Rheumatism;Diabetes
146	*Olax scandens*	Olacaceae	Climber	Fevers;Anaemia
147	*Opiliaamentacea*	Opilaceae	Climber	Snake bite
148	*Bamboosaarundinaceae*	Bambusiaceae	Bush	Leprosy;Wounds
149	*Bauhinia vahlii*Wight&Arn	Caesalpiniaceae	Creeper	Aphrodisiac;Dysentery
150	*Dioscorea pentaphylla* L.	Dioscoreacea	Vine	Leprosy;Dysentery

CONCLUSION

Medicinal plants have been discovered and used in traditional medicine practices since prehistoric times also hundreds of chemical compounds for functions including defence against insects, fungi, diseases, and herbivorous mammals were known since then. Somany medicinal plants are resources of new drugs and many of the modern medicines are produced indirectly from the plants. The medicinal plants helps to protect human and animals from natural poisons. The importance of traditional medicine has also recognized by WHO.

REFERENCES

1. Chattopadhyay RR and Bhattacharrya SK. Plant Review *Terminalia chebula*: An update. Pharmacognosy Reviews Journal, 2007, 1(1): 151-155.
2. Gowda DV, Muguli Ganesh, Rangesh PR and Deshpande Rohan D. Phytochemical and Pharmacological actions of Triphala: Ayurvedic Formulation - A Review. International Journal of Pharmaceutical Sciences Review & Resear, 2012, 15(2): 61-65.
3. Amit Kumar. A review of traditional anticancer nano-medicine: Triphala. The Journal of Pharma Innovation, 2014, 3(7): 60-66.
4. Padmaa Paarakh. *Terminalia arjuna* (Roxb.) Wt.and Arn. : A Review. International Journal of Pharmacology, 2010, 5(5): 1-5.
5. Bariya Aditi Hemrajbhai, Patel Madhabhai Manohardas, Anand Indermeet Singh, Darji Vinay Chhanalal. A Phytochemical and Pharmacological review on *Terminalia arjuna*, Pharma Research Library Journal, 2016, 4(2): 1-5.
6. Anindita Deb, Sikha Barua, Dr. Biswajit Das. Pharmacological activities of Baheda (*Terminalia bellerica*): A review, Journal of Pharmacognosy and Phytochemistry 2016; 5(1): 194-197
7. Ravi Kant Upadhyay. Nutraceutical, pharmaceutical and therapeutic uses of Allium cepa: A review. International Journal of Green Pharmacy. 2016; 10 (1) : 46-64.
8. Guo-Qing Shi, Jing Yang, Jiang Liu, Sheng-Nan Liu1, Han-Xue Song, Wen-En Zhao and Yan-Qi Liu. Isolation of flavonoids from onion skin and their effects on K562 cell viability. A Journal of the Bangladesh Pharmacological Society. 2016; 11:18-25.
9. Olanrewaju Sam Olayeriju, Mary Tolulope Olaleye, Olamide Olajusi Crown, Kayode Komolafe, Aline Augusti Boligon, Margareth Linde Athayde and Akintunde Afolabi

Akindahunsi. Ethylacetate extract of red onion (Allium cepa L.) tunic affects hemodynamic parameters in rats. 2015, 4(3):115-122.

10. Lee J and Mitchell AE. Quercetin and Isorhamnetin glycosides in onion; Varietal comparison, physical distribution, coproduct evaluation and long term storage stability. Journal of Agricultural and Food Chemistry. 2011; 59(3): 857-63.

11. Praveen Kumar A, Bhawana S. 2013. HPLC Analysis and Isolation of Rutin from Stem Bark of Ginkgo biloba L. Journal of Pharmacognosy and Phytochemistry 2(4):68-71.

12. Maria Isabel Cardona, Reina Marcela Toro, Geison M. Costa, Luis Fernando Ospina, Leonardo Castellanos, Freddy A. Ramos, Diana Marcela Aragón. 2017. Influence of extraction process on antioxidant activity and rutin content in Physalis peruviana calyces extract. Journal of Applied Pharmaceutical Science, 7 (06), 164-168.

13. Fatemeh Fathiazada, Abbas Delazara, Roya Amiria, Satyajit D. 2006. Sarker. Extraction of Flavonoids and Quantification of Rutin from waste Tobacco Leaves. Iranian Journal of Pharmaceutical Research 3: 222-227.

14. Stevenson DG, Eller FJ, Wang L, Jane JL, Wang T, Inglett GE (2007). Oil and tocopherol content and composition of pumpkinseed in 12 cultivars, J Agric Food Chem, 55:400504013

15. Li T, Ito A, Chen X, Long C, Okamoto M, Raoul F, Giraudoux P,Yanagida T, Nakao M, Sako Y, Xiao N, Craig PS (2012). Use-fulness of pumpkin seeds combined with areca nut extract incommunity-based treatment of human taeniasis in northwestSichuan Province, China. Acta Trop 124:1520157.

16. Caili F, Huan S, Quanhong L (2006). A review on pharmacological activities and utilization technologies of pumpkin, Plant Foods Hum Nutr, 61: 73-80.

17. Fu C, Shi H, Li Q (2006). A review on pharmacological activities and utilization technologies of pumpkin, J Plant Foods Hum Nutr, 61(2):73-80.

18. Anwesa Bag, Subir Kumar Bhattacharyya, Rabi Ranjan Chattopadhyay. The development of Terminalia chebula Retz. (Combretaceae) in clinical research. Asian Pacific Journal of Tropical Biomedicine, 2013; 3(3): 244-252.

19. Mishra RN. Rasayan – The Ayurvedic Perspective. Research Journal of Pharmaceutical, Biological and Chemical Sciences, 2011, 2(4), 269-282.

20. S.K Sharma, Lalit Singh, Suruchi Singh. A Review On Medicinal Plants Having Antioxidant Potential. Indian Journal of Research in Pharmacy and Biotechnology, 2013, 1(3), 404-409.

21. Prasan R. Bhandari, Mohammad Ameeruddin Kamdod. Emblica officinalis (Amla): A review of potential therapeutic applications. International Journal of Green Pharmacy, 2012, 6(4), 257-269.

22. S.S. Patel and R.K. Goyal. Emblica officinalis Geart.: A Comprehensive Review on Phytochemistry, Pharmacology and Ethnomedicinal Uses, Research Journal of Medicinal Plant, 2012, 6: 6-16.

23. V. Eugin Amala, M. Jeyaraj. Hepato Protective Efficacy of Terminalia chebula, Terminalia bellirica, Phyllanthus emblica and Their Formulation on Imidacloprid Induced Liver Toxicity by Histopathological and Biochemical Parameters. International Journal of Pharmaceutical Sciences and Drug Research 2015; 7(5): 407-411.

24. Bao Yang, Yueming Jiang, John Shi, Feng Chen, Muhammad Ashraf. Extraction and pharmacological properties of bioactive compounds from longan (Dimocarpus longan Lour.) fruit — A review. Food Research International Journal, 2011, 44:1837–1842.

25. Naveen Srinivas, Ketki Sali, Atul A Bajoria· Therapeutic aspects of Tulsi unraveled: A review. Journal Of Indian Academy Of Oral Medicine and Radiology, 2016, 28: 17-23

26. U Anyaehie. Medicinal properties of fractionated acetone/water neem (Azadirachta indica) leaf extract from Nigeria: a review, nigerian journal of physiological sciences 2009, 24(2):12-17.

27. Douglas Grindlay, T.Reynolds The *Aloe vera* phenomenon: A review of the properties and modern uses of the leaf parenchyma gel, journal of ethnopharmacology, 2004, 16:2-3.

28. N.Matan, H.Rimkeeree, A.J.Mawson, P.Chompreeda, V.Haruthaithanasan, M.Parker. Antimicrobial activity of cinnamon and clove oils under modified atmosphere conditions, international journal of food microbiology, 2006, 107(2):180- 185.

29. Shrikant Mishra and Kalpana Palanivelu. The effect of curcumin (turmeric) on Alzheimer's disease, Journal of Indian academy of neurology, 2008,11(1): 1-10.

30. M.D., Ph.D.Benjamin, H.S.Lau, M.A.Moses A.Adetumbi, Dr.P.H.AlbertSanchez. Allium sativum (garlic) and atherosclerosis: A review, nutritional research journal, 1983, 3(1):119-128.